MARS

MARYSA STORM

BLACK RABBIT BOOKS

Bolt Jr. is published by Black Rabbit Books
P.O. Box 227, Mankato, Minnesota, 56002.
www.blackrabbitbooks.com

Michael Sellner, designer and photo researcher

Names: Storm, Marysa, author.
Title: Mars / by Marysa Storm.
Description: Mankato, Minnesota : Black Rabbit Books, [2023]
Series: Bolt Jr. space explorer | Includes bibliographical references and index. | Audience: Ages 6-8 | Audience: Grades K-1 | Summary: "Suit up! Mars is a planet like no other, and it's time to explore it through a colorful design that features vibrant photography, engaging infographics, and carefully leveled text"– Provided by publisher.
Identifiers: LCCN 2021026224 (print) | LCCN 2021026225 (ebook) | ISBN 9781623107741 (hardcover) | ISBN 9781644665954 (paperback) | ISBN 9781623107826 (ebook)
Subjects: LCSH: Mars (Planet)–Juvenile literature.
Classification: LCC QB641 .S67 2023 (print) | LCC QB641 (ebook) | DDC 523.43–dc23
LC record available at https://lccn.loc.gov/2021026224
LC ebook record available at https://lccn.loc.gov/2021026225

Image Credits

123RF: pike28, 22–23; Alamy: Thibault Renard, 6–7; UPI, 18–19; Getty: MARK GARLICK/SCIENCE PHOTO LIBRARY, 6–7; iStock: JoeLena, 20–21; Science Source: DETLEV VAN RAVENSWAAY, 16–17; ESO/Martin Kornmesser/Nick Risinger, 4–5; MARK GARLICK, 4–5, 10–11; Steven Hobbs/Stocktrek Images, 8–9; Shutterstock: Bikomins, 3, 24; buradaki, 12; Dima Zel, Cover; fukume, 1; Humannet, 20–21; Mopic, 14–15; Sineenat, 12–13; Vadim Sadovski, 12–13; Vectorfair.com, 14–15; Vector Tradition, 10–11

Contents

CHAPTER 1

Meet the Planet

Billions of years ago, Mars was warm. Oceans might have covered its **surface**. It might have even had life. The planet looks much different now.

surface: the outside of an object

COMPARING SIZES

Mars
13,233 miles
(21,297 kilometers)
around

Dry and Dusty

Now, Mars is dry. It's dusty. Instead of oceans, it has wide holes called craters. Life cannot survive on it.

Features

plains
no rings
craters

CHAPTER 2

Its Place in Space

There are four **inner planets**. Earth is one of them. So is Mars. Like other inner planets, it's rocky.

inner planet: any of the planets that lie between the sun and the asteroid belt

FACT

Mars has dust storms.

Hot and Cold

Mars is the fourth planet from the sun. It gets warm during the day. But the planet has a thin **atmosphere**. It can't hold in heat. It drops to -284 degrees Fahrenheit (-175.6 degrees Celsius) at night.

atmosphere: the gases that surround a planet

Mars in the Solar System

Neptune
Uranus
Saturn
Jupiter
asteroid belt

CHAPTER 3

Explore!

Many machines have visited Mars. The first rover landed on it in 1997. In 2021, a helicopter flew on Mars. It was the first machine to take off from the planet.

FACT

Rovers take pictures and gather info.

The Future

Robots continue studying Mars. People hope to visit the planet too. Someday, they might live there.

the rover and its helicopter that landed in 2021

Perseverance and Ingenuity

Bonus Facts

People call it the Red Planet.

Mars has two moons.

Its days last about 25 hours.

Its color comes from rust.

rust: a reddish coating formed on metal

READ MORE/WEBSITES

Finan, Catherine C. *Planets and Moons.* Space. Minneapolis: Bearport Publishing, 2022.

Grack, Rachel. *Curious about Mars.* Curious About. Mankato, MN: Amicus/Amicus Ink, 2022.

Leed, Percy. *Mars: A First Look.* Read about Space. Minneapolis: Lerner Publications, 2023.

Astronomy: Planet Mars
www.ducksters.com/science/mars.php

Mission to Mars
kids.nationalgeographic.com/space/article/mission-to-mars

GLOSSARY

atmosphere (AT-muhs-feer)—the gases that surround a planet

inner planet (IN-er PLAN-it)—any of the planets that lie between the sun and the asteroid belt

rust (RUHST)—a reddish coating formed on metal

surface (SUR-fus)—the outside of an object

INDEX